THE POETRY OF ZIRCONIUM

The Poetry of Zirconium

Walter the Educator™

SKB

Silent King Books a WhichHead Imprint

Copyright © 2023 by Walter the Educator™

All rights reserved. No part of this book may be reproduced in any manner whatsoever without written permission except in the case of brief quotations embodied in critical articles and reviews.

First Printing, 2023

Disclaimer
This book is a literary work; poems are not about specific persons, locations, situations, and/or circumstances unless mentioned in a historical context. This book is for entertainment and informational purposes only. The author and publisher offer this information without warranties expressed or implied. No matter the grounds, neither the author nor the publisher will be accountable for any losses, injuries, or other damages caused by the reader's use of this book. The use of this book acknowledges an understanding and acceptance of this disclaimer.

"Earning a degree in chemistry changed my life!"
- Walter the Educator

dedicated to all the chemistry lovers, like myself, across the world

CONTENTS

Dedication v

Why I Created This Book? 1

One - Oh, Zirconium 2

Two - Element Divine 4

Three - Strength And Love 6

Four - Darkest Night 8

Five - Conductor Of Heat And Power . . . 10

Six - Gemstone Of Science 12

Seven - Artistic Flair 14

Eight - Exquisite Heart 16

Nine - Sparks Inspiration 18

Ten - Day And Night 20

Eleven - Zirconium, A Companion 22

Twelve - Magic And Power 24

Thirteen - Tale Of Grandeur	. . . .	26
Fourteen - Reverence And Awe	. . . .	28
Fifteen - Enduring Pleasure	. . . .	30
Sixteen - Minds Explore	. . . .	32
Seventeen - Nuclear Reactors	. . . .	34
Eighteen - Metal Of Wonder	. . . .	36
Nineteen - Bursting At The Seams	. . . .	38
Twenty - Hour By Hour	. . . .	40
Twenty-One - Luminary	. . . .	42
Twenty-Two - Marvel To Chase	. . . .	44
Twenty-Three - Gift From Above	. . . .	46
Twenty-Four - Silent Guardian	. . . .	48
Twenty-Five - Transforming Into Beauty	. . .	50
Twenty-Six - Industry And Science	. . . . . .	52
Twenty-Seven - Cornerstone Of Progress	. .	54
Twenty-Eight - Jewel Of Science	. . . .	56
Twenty-Nine - Wondrous Fable	. . . .	58
Thirty - Pillar Of Strength	. . . .	60
Thirty-One - Mystical Prime	. . . .	62
Thirty-Two - Cosmic Traveler	. . . .	64

Thirty-Three - A Masterpiece 66

Thirty-Four - Scientific Delight 68

Thirty-Five - Catalyst Of Change 70

About The Author 72

WHY I CREATED THIS BOOK?

Creating a poetry book about the chemical element of Zirconium was an intriguing and unique endeavor. Zirconium, with its fascinating properties and applications, offered a rich source of inspiration. By exploring the various aspects of Zirconium, such as its atomic structure, history, uses, and symbolism, I can delve into themes of transformation, resilience, beauty, and the interconnectedness of science and art. This poetry book can serve as an educational tool, captivating readers with the beauty and elegance of poetry while imparting knowledge about Zirconium. It can also be a means to showcase the creativity and versatility of poetry by exploring unusual and unexpected subjects. Overall, creating a poetry book about Zirconium presents an opportunity to merge science and art, infusing the seemingly mundane with beauty and meaning.

ONE

OH, ZIRCONIUM

In the heart of Earth's ancient core,
A jewel of strength, forevermore.
Zirconium, a radiant flame,
A catalyst in nature's game.
    A metal strong, yet light and pure,
A luminary, steadfast and sure.
In crystal form, a precious sight,
With luster gleaming, shining bright.
    A guardian of heat, it stands,
Withstanding fires from distant lands.
In nuclear power, it finds its place,
A shield against the cosmos' embrace.
    From oceans deep to mountains high,
Zirconium's presence does imply,

A quiet force, a hidden gem,
Embedded in Earth's ancient realm.
 In laboratories, it finds its use,
A catalyst to progress, to infuse,
In chemical reactions, it plays its part,
Guiding transformations, a work of art.
 Oh, Zirconium, noble and true,
A symbol of resilience, through and through,
In the realm of elements, you shine,
A testament to nature's grand design.
 So let us honor this precious metal,
With respect and awe, our hearts settle,
For in its essence, we find a story,
Of strength, beauty, and ancient glory.

TWO

ELEMENT DIVINE

In the realm of elements, a gem unseen,
Zirconium shines with a radiant gleam.
A metal so noble, a jewel so rare,
It carries a secret, a tale to share.

    Born in the depths of Earth's fiery core,
Zirconium emerges, forever to adore.
With atomic strength and a lustrous hue,
Its presence enthralls, its beauty so true.

    A silent guardian, steadfast and strong,
Zirconium's virtues we've known all along.
A protector of metals, it shields and defends,
Against corrosion, it courageously contends.

    In laboratories, it sparks innovation's fire,
A catalyst for progress, never to tire.

With boundless potential, it takes new form,
Transforming ideas and weathering the storm.

Zirconium, the conductor of dreams,
Unleashing power, or so it seems.
From nuclear reactors to jewelry so fine,
Its versatility transcends the confines.

From Earth's ancient rocks to distant space,
Zirconium's legacy we proudly embrace.
A symbol of resilience, its story untold,
In every atom, a secret to behold.

So let us celebrate this element divine,
Zirconium, a jewel that forever will shine.
A testament to nature's grand design,
In its presence, a brilliance we forever find.

THREE

STRENGTH AND LOVE

In the depths of the earth, where fires burn bright,
Lies a treasure unseen, a radiant light.
Zirconium, oh element of grace,
In your atomic embrace, secrets we chase.

    A lustrous metal, silvery and strong,
You dance with beauty, you sing a song.
From the crust to the stars, your journey unfolds,
A story untold, as time gently molds.

    In the core of a star, you were born,
Through cosmic storms and celestial scorn.
From the heart of a supernova's demise,
You emerged, a gem to mesmerize.

    In the lab of the alchemist's dream,
Your properties gleam, a magical beam.

Resilient and robust, you stand tall,
In alloys and ceramics, you conquer all.
    Zirconium, the artist's delight,
You paint the canvas with colors so bright.
In jewelry and gemstones, you shine with pride,
A symbol of love, forever tied.
    And in the nuclear reactor's core,
You harness the power, forevermore.
With control and precision, you light the way,
A beacon of progress, for night and day.
    Zirconium, element of wonder and might,
Forever you'll guide us, like a guiding light.
From deep beneath the earth to the heavens above,
You teach us the power of strength and love.

FOUR

DARKEST NIGHT

In the depths of the Earth's embrace,
Lies a treasure of strength and grace.
Zirconium, noble and serene,
A metal rare, a beauty unseen.

From the stars it came, in fiery birth,
A cosmic dance, a moment of worth.
Bound by nature's alchemical hand,
Zirconium, a gem of the land.

A lustrous sheen, like moonlit mist,
It whispers secrets through a gentle twist.
Its atomic heart, steadfast and true,
A guardian of dreams, both old and new.

In the crucible of time, it stood,
A symbol of resilience, misunderstood.

For in its core, a power untold,
A testament to the stories it holds.
    Zirconium, a bridge to the unknown,
A conductor of light, a pathway shown.
Its presence, a reminder to be aware,
Of the wonders that lie beyond our stare.
    So let us cherish this element rare,
For it embodies the strength we all share.
Zirconium, a beacon of hope and might,
Guiding us through the darkest night.

FIVE

CONDUCTOR OF HEAT AND POWER

In the heart of Earth's fiery core,
Where molten rivers churn and roar,
Lies a gem, a secret rare,
Zirconium, beyond compare.

A gleaming metal, strong and true,
Its beauty hidden from our view,
A silvery shade, a lustrous hue,
Zirconium, with mysteries imbued.

From ancient stars, it was born,
Forged in cosmic fires, we adorn,
A noble element, steadfast and pure,
Zirconium, forever to endure.

In laboratories, it finds its way,
A catalyst, guiding reactions' sway,

A bridge between the atoms' dance,
Zirconium, enhancing life's advance.

In the quiet depths of night,
It glows with phosphorescent light,
A beacon in the darkness deep,
Zirconium, a secret it does keep.

A conductor of heat and power,
Zirconium, in every hour,
In reactors, it takes its stand,
Powering cities, across the land.

And in the realm of jewelry's grace,
Zirconium finds a sacred place,
Adorning fingers, necks, and ears,
Zirconium, a symbol of love that endears.

Oh, Zirconium, element divine,
In your essence, we find a sign,
Of strength, of beauty, of secrets rare,
Zirconium, in you, we share.

SIX

GEMSTONE OF SCIENCE

In the realm of elements, a jewel does shine,
Zirconium, a treasure, so rare and divine.
With atomic number forty, it claims its place,
A metal of wonder, with elegance and grace.

Its lustrous allure, a silvery-white sheen,
Zirconium, the element, a beauty unseen.
Strong and resilient, it stands the test of time,
A symbol of endurance, a reason to rhyme.

In the depths of the Earth, where secrets reside,
Zirconium emerges, a gem to confide.
From minerals and ores, it's carefully obtained,
A gemstone of science, its wonders unchained.

With heat and precision, it's crafted with care,
Zirconium, the jewel, that's beyond compare.

In reactors it thrives, as fuel rods ignite,
Unleashing its power, a celestial light.

As jewelry adorns, in rings and in bands,
Zirconium enchants, with its magical hands.
A symbol of commitment, love's eternal flame,
Forever entwined, in life's glorious game.

So let us rejoice, in the element's might,
Zirconium, the gem, that shines ever so bright.
A testament to science, and nature's grand plan,
Zirconium, the treasure, that captivates every fan.

SEVEN

ARTISTIC FLAIR

In the realm of elements, a wonder resides,
Zirconium, a marvel that nature provides.
A metal, steadfast, with a gleaming grace,
Unveiling secrets in its atomic embrace.

A symbol of strength, it stands tall and true,
With properties rare, known to just a few.
From Earth's deep core, it emerges bright,
A beacon of light, in the darkest of night.

Zirconium, a guardian of fire's might,
Igniting sparks with its incandescent light.
A conductor of heat, it embraces the flame,
Unyielding and resolute, in its vibrant game.

In its crystalline form, it glimmers and shines,
A prism of colors, a treasure that defines.

Its beauty unmatched, like a gemstone rare,
A testament to nature's artistic flair.

Yet, Zirconium's allure goes far beyond sight,
In nuclear realms, it takes on a new height.
A shield against chaos, a fortress so strong,
Protecting the world, as it moves along.

From ancient civilizations to modern-day,
Zirconium's presence continues to sway.
A versatile element, with a story untold,
Creating marvels, as its secrets unfold.

Oh, Zirconium, element of wonder and awe,
In your atomic dance, we stand in awe.
Forever we'll cherish your celestial dance,
Zirconium, a gem in nature's vast expanse.

EIGHT

EXQUISITE HEART

In the depths of earth, a treasure gleams,
A metal rare, born from ancient dreams.
Zirconium, a shimmering delight,
Unveiling secrets, casting radiant light.

A noble element, strong and true,
With atomic prowess, it breaks through.
In fiery furnaces, it takes its form,
Transforming sand into beauty, reborn.

Its crystal lattice, structured and pure,
Endows zirconia with strength to endure.
A gemstone of wonder, dazzling and bright,
Reflecting the stars with celestial might.

In nuclear reactors, it finds its place,
Harnessing energy with steady grace.

A guardian of power, a sentinel strong,
Guiding the atoms, where they belong.

    A conductor of heat, a warrior bold,
Zirconium's touch, never grows cold.
In engines and turbines, it dances with flame,
Pushing the limits, igniting the game.

    Yet, beyond its strength, beyond its might,
Lies a beauty that captivates sight.
In the eyes of the beholder, a delicate hue,
Irresistible charm, forever true.

    Zirconium, oh wondrous element divine,
Your existence, a marvel, so fine.
A symbol of endurance, a symbol of art,
A testament to nature's exquisite heart.

NINE

SPARKS INSPIRATION

In the depths of the earth, a hidden gem so rare,
Zirconium, a metal, beyond compare.
A radiant presence, with secrets untold,
A tale of strength and beauty, waiting to unfold.
    Born from the stars, in fiery cosmic birth,
Zirconium emerges, a symbol of worth.
Its lustrous sheen, a shimmering delight,
Casting a spell, captivating every sight.
    A guardian of time, it stands firm and strong,
Resilient and steadfast, where it belongs.
With atomic prowess, it binds and combines,
Creating alloys, where innovation shines.
    Zirconium, a catalyst, a catalyst for change,
A catalyst for progress, it rearranges.

From the depths of its core, it sparks inspiration,
Igniting creativity, a divine revelation.

In laboratories, it dances with flame,
Unleashing its power, a delicate game.
Enhancing life's wonders, with a touch so fine,
Zirconium's magic, forever will shine.

In jewelry and art, it finds its repose,
Adorning brilliance, where elegance flows.
A symbol of love, forever sealed,
Zirconium's allure, forever revealed.

Oh, Zirconium, element of grace,
You leave us in awe, with your celestial embrace.
A testament to nature's grand design,
Zirconium, forever, will eternally shine.

TEN

DAY AND NIGHT

In the depths of earth's ancient core,
Where molten fires fiercely roar,
Lies a gem, a treasure untold,
A metal element, strong and bold.

Zirconium, with its gleaming hue,
A symbol of strength, forever true,
Within its atoms, secrets reside,
Unveiling a world, where wonders hide.

From the stars, it was born,
In cosmic explosions, it was torn,
Now it rests, in rocks and sands,
Waiting to be held in human hands.

A lustrous metal, shining bright,
Zirconium, a beacon of light,

With unmatched beauty, it does shine,
A gem that nature did design.

Its presence brings a calming grace,
A soothing touch, a gentle embrace,
In laboratories, it finds its place,
A catalyst, advancing our race.

Zirconium, a bridge to the new,
With alloys strong, it helps us through,
From nuclear reactors to jet engines' might,
It powers our dreams, day and night.

Oh, noble Zirconium, element divine,
In your essence, we find solace and rhyme,
A symbol of progress, wisdom, and might,
Guiding us towards a future so bright.

So let us cherish this element rare,
With gratitude, awe, and utmost care,
For Zirconium, in its shimmering sheen,
Embodies the wonders of what's unseen.

ELEVEN

ZIRCONIUM, A COMPANION

In the realm of the periodic table's dance,
Zirconium stands with a dignified stance.
A metal of strength, in its atomic core,
A tale of wonder, forevermore.

A shining gem born in the celestial fire,
Zirconium's allure never seems to tire.
With a gleam that rivals the brightest star,
It enchants us all, from near and far.

A guardian of bridges, towers, and spires,
Zirconium withstands nature's fierce desires.
Its resilience unmatched, a true marvel to see,
A testament to its durability.

Within its structure, secrets lie deep,
A crystal lattice, a treasure to keep.

From nuclear reactors to spacecraft's frame,
Zirconium's presence seals their fame.
   Its blends with alloys, a symphony in tune,
Strong and corrosion-resistant, a boon.
A conductor of heat, a conductor of light,
Zirconium shines through the darkest night.
   But beneath its strength, a gentle soul resides,
A silvery-white beauty, love confides.
A silent companion, ever by our side,
Zirconium, a companion, our hearts can't hide.
   So, let us raise a toast to this noble element,
Zirconium, forever in our sentiment.
A symbol of strength, beauty, and grace,
In the realm of chemistry, it finds its place.

TWELVE

MAGIC AND POWER

In the realm of elements, a jewel unseen,
Zirconium, a shimmering dream.
A metal of strength, with secrets untold,
Its story, a treasure yet to unfold.
 Born in the stars, in celestial fires,
Zirconium, the cosmic desires.
Through eons it traveled, in stardust's embrace,
To Earth's fertile soils, finding its place.
 A lustrous beauty, with a silver sheen,
Zirconium, a rare and radiant gleam.
In crystal lattice, its atoms align,
A symphony of structure, harmonious and fine.
 It dances with light, a prism of hues,
Zirconium, the enchantment it brews.

Unyielding and steadfast, it stands the test,
Against time's grip, it remains at its best.

From jet engines soaring, to bridges that span,
Zirconium, the strength of a steadfast plan.
In medical marvels, it lends a hand,
Implants of grace, a life's second chance.

A symbol of wisdom, a beacon of truth,
Zirconium, a legacy of youth.
For in its core, a fire burns bright,
Igniting the passions, igniting the fight.

Zirconium, a tale yet to be told,
A metal of wonder, a story so bold.
As we unravel its mysteries deep,
Its magic and power, forever we'll keep.

THIRTEEN

TALE OF GRANDEUR

In the realm of elements, a gem doth reside,
With strength and allure, its essence does hide.
Zirconium, the element, rare and divine,
A tale of grandeur, let me now define.

    A lustrous metal, with a silver sheen,
Zirconium sparkles, a radiant beam.
Its atomic number, forty and two,
A symbol of power, steadfast and true.

    From the Earth's crust, it does emerge,
A precious gem, nature's splendid surge.
In crystal form, Zirconium gleams,
A testament to its magnificent dreams.

    With strength unyielding, it withstands the heat,
A champion of fire, it won't retreat.

In nuclear reactors, it finds its place,
A guardian of power, a cosmic embrace.
   Zirconium, oh noble and grand,
A symbol of courage, across the land.
Its properties diverse, like a chameleon's hue,
An element so wondrous, so vibrant and true.
   So let us honor this gem of the Earth,
A symbol of beauty, of infinite worth.
Zirconium, we raise our voice to thee,
Forever admired, for all to see.

FOURTEEN

REVERENCE AND AWE

In realms unseen, where wonders lie,
A gleaming element catches the eye.
Zirconium, a jewel of the Earth,
Birthed in ancient fires, of celestial birth.

A metal rare, yet unassuming,
Its secrets hidden, ever blooming.
A lustrous gem, with a touch of grace,
Unraveling mysteries, at its own pace.

From molten depths, its essence emerged,
A radiant spirit, with strength submerged.
A guardian of dreams, it quietly gleams,
A bridge between realms, beyond our schemes.

With atomic might, it binds and weaves,
A tapestry of life, from roots to leaves.

In bones and shells, it finds its place,
A silent architect, in nature's embrace.

From stars above, it traveled afar,
A cosmic wanderer, a shining star.
With every heartbeat, it pulses strong,
A symphony of atoms, a timeless song.

Zirconium, a shimmering gem,
A testament to the Earth's diadem.
In its embrace, we find connection,
A reminder of our shared inception.

So let us honor this element divine,
With reverence and awe, let us intertwine.
For in the depths, where secrets lie,
Zirconium whispers, with a silent sigh.

FIFTEEN

ENDURING PLEASURE

In a world of elements, precious and rare,
There lies a beauty, beyond compare.
Zirconium, a jewel in the Earth's core,
Radiates brilliance, forevermore.

With atomic number forty, it resides,
In the periodic table, where it hides.
A lustrous metal, strong and true,
Zirconium, I sing this ode to you.

Born in the heart of a blazing star,
Forged in cosmic fires, traveling far.
From the depths of time, you emerged,
A symbol of strength, your power surged.

Your presence, unyielding, steadfast and bold,
In alloys and ceramics, your tales unfold.

A catalyst of change, a catalyst of dreams,
Zirconium, the element that gleams.
In hues of silver, you shimmer and shine,
Reflecting the light, so divine.
A symbol of purity, flawless and clear,
Zirconium, you hold secrets so dear.
From jewelry to nuclear reactors, you find your place,
Safeguarding our world, with grace.
A guardian of progress, an element of might,
Zirconium, you shine through the night.
Oh, Zirconium, with your atomic might,
You captivate us, day and night.
A marvel of science, a wonder to behold,
Zirconium, your story, forever untold.
So let us celebrate this precious treasure,
Zirconium, a symbol of enduring pleasure.
In the realm of elements, you stand apart,
Zirconium, forever etched in our heart.

SIXTEEN

MINDS EXPLORE

In the depths of Earth's ancient core,
A treasure lies, forevermore.
Zirconium, a secret unbound,
A metal rare, profound.

Born in the fiery heart of stars,
Forged in celestial fires afar,
Zirconium, a cosmic gem,
Shines bright, its mystic stem.

A lustrous beauty, silver-white,
Unveiling secrets, day and night.
Its atomic dance, a silent song,
Infinite possibilities, lifelong.

With strength and grace, it takes its place,
An element of power and embrace.

Zirconium, a sturdy shield,
Against nature's forces, it will wield.
    In laboratories, minds explore,
Zirconium, a key to open doors.
Unleashing brilliance, boundless might,
A catalyst in science's flight.
    From jet engines soaring high,
To implants that heal, never say goodbye.
Zirconium, a friend to man,
Aiding progress, with a helping hand.
    In jewelry's embrace, it finds its art,
Adorning hands and seizing hearts.
Zirconium, a symbol of love,
A reminder, from Earth to above.
    So let us cherish, this element rare,
Zirconium, beyond compare.
For in its essence, we can find,
The wonders of creation, intertwined.

SEVENTEEN

NUCLEAR REACTORS

In the realm of elements, Zirconium gleams,
A lustrous metal that captivates dreams,
Its atomic number, forty, stands tall,
A symbol of strength, it does proudly install.

Born in the core of a blazing star's birth,
Zirconium dances upon this vast Earth,
With a silvery sheen and a radiant glow,
It weaves tales of wonder, for all to bestow.

A guardian of secrets, it guards with its might,
In alloys and compounds, it finds its true light,
Sturdy and robust, it endures the test,
Zirconium, the element that outshines the rest.

In nuclear reactors, it finds its own grace,
Harnessing energy, a celestial embrace,

Its isotopes stable, they stand firm and true,
A beacon of hope, a promise anew.
    Zirconium, the element, a jewel in its core,
Its beauty and strength leave us wanting for more,
With its intricate patterns, enchanting and rare,
It captures our hearts, an element so fair.
    So let us embrace this wondrous metal,
With admiration and awe, let us settle,
For Zirconium, a marvel, a treasure untold,
A symbol of resilience, forever bold.

EIGHTEEN

METAL OF WONDER

In the realm of elements, Zirconium shines bright,
A metal of wonder, a captivating light.
With atomic number forty, it claims its place,
Unveiling its secrets, with elegance and grace.

From the depths of the Earth, it emerges, refined,
A jewel of creation, a treasure to find.
Its lustrous white hue, like the moon's gentle glow,
Zirconium, a marvel, in its cosmic show.

Beneath the surface, where heat and pressure abide,
Zirconium's story is etched, deep inside.
Crystals of beauty, with patterns so rare,
A testament to nature's artistry, beyond compare.

Steadfast and resilient, it stands the test of time,
A guardian of strength, in a world that's filled with grime.

In alloys and ceramics, it finds its true worth,
Enhancing their prowess, bringing value and mirth.

In nuclear reactors, it plays a vital role,
Harnessing energy, with a purposeful goal.
A guardian of power, in this modern age,
Zirconium, the keeper of a nuclear stage.

From jewelry to industry, it lends its might,
Zirconium, a symbol of innovation's flight.
A silent hero, silently it thrives,
Unassuming and humble, as it constantly strives.

So let us raise a toast, to this element divine,
Zirconium, a jewel in the periodic line.
A testament to science, to discovery's quest,
Zirconium, forever in our hearts, we'll nest.

NINETEEN

BURSTING AT THE SEAMS

In the heart of Earth's hidden embrace,
Lies a treasure that glistens with grace.
Zirconium, a gem of elements rare,
With secrets and wonders beyond compare.

A radiant metal with a lustrous sheen,
Zirconium, in nature's grandest scene.
A chameleon, shifting in shades,
From silver to gray, its beauty cascades.

Within its core, a strength untamed,
As fire and heat it fearlessly claimed.
A guardian of reactors, it stands tall,
Harnessing power, for one and all.

From the depths of Earth, it emerges bright,
A beacon of hope, a guiding light.

With unyielding resolve, it never wanes,
Zirconium, the element that sustains.

In jewelry and art, its presence adorns,
An emblem of elegance, a crown of thorns.
A symbol of endurance, it reminds,
That life's challenges, we can always bind.

Zirconium, a guardian of dreams,
Unveiling mysteries, bursting at the seams.
In laboratories, it dances with fire,
Pushing boundaries, reaching higher.

So let us celebrate this element true,
Zirconium, we honor you.
For in your essence, we find delight,
A testament to nature's endless might.

TWENTY

HOUR BY HOUR

In the depths of Earth's embrace, lies Zirconium's hidden grace,
A metal of strength and timeless allure, a marvel to endure.
Born amidst the cosmic fire, forged in the heart of ancient pyre,
Zirconium, a radiant star, gleaming from afar.

Its atomic dance, a symphony of might, a dance that conquers the darkest night,
With valence electrons, it weaves a spell, a tale of magic none can quell.
From molten depths, it rises high, a phoenix soaring through the sky,
Zirconium, a beacon of light, shimmering in the darkest plight.

In laboratories, minds explore, unlocking secrets at its core,
A catalyst of wondrous power, shaping molecules, hour by hour.
In alloys, it finds its place, enhancing strength with every trace,
Zirconium, a guardian bold, protecting treasures yet untold.

From jewelry adorning a lover's neck, to nuclear reactors, a delicate check,
Zirconium, versatile and pure, a testament to science's allure.
Its brilliance captured in every gem, a testament to Earth's rarest stem,
Zirconium, a gemstone divine, a treasure for those who seek to find.

So let us honor this element grand, within our hearts, let it expand,
Zirconium, a gift from nature's hand, a symbol of strength, forever to stand.
In the realm of elements, it shines unique, a jewel of the periodic clique,
Zirconium, a radiant gem, a marvel we can't help but condemn.

TWENTY-ONE

LUMINARY

In the realm of elements, behold Zirconium's grace,
A metal of grandeur, found in nature's embrace.
With atomic number forty, it proudly stands,
A symbol of strength, forged by celestial hands.

From Earth's crust it arises, buried deep within,
A treasure sought by those who yearn to begin.
Its lustrous appearance, a shimmering delight,
Captivating whispers in the darkest of night.

Zirconium, oh noble friend, with beauty so rare,
A shining beacon, beyond compare.
Its strength enshrined within its core,
Resilient and steadfast, forevermore.

In alloys it dances, with steel it combines,
Creating a bond that forever defines,

The strength of foundations, the structures we raise,
A testament to Zirconium's unwavering blaze.

    In nuclear reactors, it plays a vital role,
Harnessing energy, a power to extol.
Radiant guardian, with purpose so clear,
Guiding humanity, dispelling all fear.

    Zirconium, a luminary in the periodic chart,
A luminescent jewel, eternally set apart.
Celestial marvel, in this cosmic ballet,
Your essence, Zirconium, forever shall stay.

    Oh, Zirconium, elemental star,
Your radiance transcends, near and far.
In the realm of elements, you reign supreme,
A testament to nature's grandest dream.

TWENTY-TWO

MARVEL TO CHASE

In a realm of wonders, a gem of grace resides,
Zirconium, a precious element, with secrets it hides.
Its lustrous sheen, like moonlight on a calm sea,
Captivates the curious, beckons them to see.

A metal of strength, steadfast and true,
Zirconium, a marvel, with powers anew.
Its atomic beauty, a shimmering dance,
Unveils the mysteries of creation, by chance.

From Earth's core, it emerges with might,
Forged in fire, bathed in celestial light.
Resilient and rare, a jewel of the Earth,
Zirconium, a testament to its noble birth.

In laboratories, scientists ponder and explore,
Unlocking Zirconium's essence, forevermore.
Its properties, versatile, a marvel to behold,
Catalyst of change, in stories yet untold.

In the depths of time, it stands steadfast,
A symbol of endurance, destined to last.
Zirconium, a guardian of the elements' might,
Guiding humanity toward a future so bright.

Oh, Zirconium, a gemstone of the skies,
With every discovery, our wonder it ties.
A tribute to the elements, a cosmic embrace,
Zirconium, forever, a marvel to chase.

TWENTY-THREE

GIFT FROM ABOVE

In the depths of the Earth, a treasure lies,
A shimmering element, Zirconium, it implies.
Its essence pure, with beauty untold,
A story to be unveiled, yet to unfold.

A metal so strong, with a heart of fire,
Zirconium, a spark of celestial desire.
In nature's embrace, it quietly resides,
A secret guardian, where mystery hides.

From ancient times, it has been sought,
For its radiant glow, a marvel it brought.
A jewel of the Earth, it graces our sight,
With every step closer, a journey of delight.

A symbol of strength, resilience, and might,
Zirconium shines, casting a celestial light.

Its atoms dancing, in harmony they bind,
Creating a masterpiece, for the world to find.

In laboratories, its secrets are revealed,
An ally to science, a treasure unconcealed.
With humble beginnings, from humble ore,
Zirconium's legacy, forevermore.

From nuclear reactors to shining rings,
Zirconium's presence, a melody it sings.
A metal of wonders, a gift from above,
Zirconium, an element we forever love.

So let us marvel at its radiant grace,
Zirconium, a gem of the Earth's embrace.
A symbol of progress, a testament to time,
In this cosmic journey, forever it will shine.

TWENTY-FOUR

SILENT GUARDIAN

In the depths of Earth's embrace, a hidden treasure lies,
A shimmering jewel, a metal of surprise.
Zirconium, a name that whispers through the night,
A chemical element, its secrets shining bright.

A guardian of strength, it stands tall and proud,
A symbol of endurance, beneath its silent shroud.
Its atomic number speaks of balance and grace,
A silent warrior, in the cosmic space.

With a crystal heart, its lattice structure gleams,
A tapestry of atoms, woven in dreams.
A bridge between worlds, it binds and unites,
A catalyst of change, igniting cosmic lights.

From the heart of stars, it was born in the fire,
Forged in supernovae, with cosmic desire.

Its beauty unmatched, its allure divine,
Zirconium, a treasure that forever will shine.

    In laboratories, it dances with chemicals of its kin,
Creating compounds, a symphony within.
From ceramics to jewelry, its uses are vast,
A versatile element, destined to last.

    Oh, Zirconium, a marvel of creation's hand,
A testament to nature's infinite command.
In your essence, a story of wonders untold,
A silent guardian, in a world so bold.

TWENTY-FIVE

TRANSFORMING INTO BEAUTY

In the depths of earth, where fire and rock collide,
A hidden treasure waits, where secrets reside.
Zirconium, a marvel, born from ancient flames,
A metal of power, with mysterious claims.

With strength unmatched, it stands tall and true,
A silent guardian, steadfast and blue.
Its atomic heart beats with pulsating might,
A radiant glow, a beacon in the night.

From molten magma, it emerges with grace,
A testament to time, in this cosmic race.
Its crystal lattice, an intricate design,
Reflecting light, a spectacle divine.

In laboratories, it dances with fire,
A catalyst of dreams, igniting desire.

A key to progress, unlocking new doors,
Advancing science, pushing the boundaries more.
    And in the hands of artists, it finds its voice,
Transforming into beauty, a work of choice.
A jewel, a gem, adorning the divine,
Captivating hearts, with elegance so fine.
    Zirconium, a symbol of endurance and might,
A precious element, shining oh so bright.
Let us marvel at its wonders, let us explore,
The magic it holds, forever to adore.

TWENTY-SIX

INDUSTRY AND SCIENCE

In the depths of the earth, a hidden gem lies,
A metal so noble, its brilliance defies.
Zirconium, the element, with strength untold,
A story of wonder, waiting to be told.

Born from stars, in cosmic explosions grand,
Zirconium emerged, a celestial band.
Its atomic heart, with thirty protons strong,
A symbol of stability, where it belongs.

A lustrous companion to the silvery moon,
Zirconium dances, casting shadows in tune.
Its radiance enchants, a mesmerizing hue,
A testament to beauty, both ancient and new.

From ancient civilizations to modern-day art,
Zirconium's allure, it leaves a lasting mark.

Adorning crowns and jewelry, so resplendent,
It captures our hearts, forever transcendent.

Yet zirconium's tale goes far beyond adornment,
In industry and science, it finds its embodiment.
A catalyst for progress, in reactors it thrives,
Powering our world, where energy derives.

So let us marvel at zirconium's might,
A symbol of resilience, shining day and night.
A bridge between worlds, both tangible and divine,
Zirconium, a treasure, forever will shine.

TWENTY-SEVEN

CORNERSTONE OF PROGRESS

In the depths of Earth's embrace, a treasure lies unseen,
A shimmering jewel of strength, a metal's noble sheen.
Zirconium, you are the heart, the core of mighty fire,
A guardian of secrets, with an iridescent desire.

Beneath the molten crust, where magma's rivers flow,
You quietly reside, a majestic cosmic glow.
Your atoms dance with grace, in a crystalline array,
Unyielding and unbreakable, forging destiny's way.

Zirconium, oh noble one, you bind the universe,
With a valiant spirit, and a power to disperse.
Your presence leaves a mark, in every earthly land,
From ancient pyramids to shores of golden sand.

A catalyst of change, in chemical symphony,

Your bonds are strong, yet flexible, in perfect harmony.
From ceramics to alloys, you lend your sturdy hand,
A cornerstone of progress, a foundation to withstand.

Zirconium, a silent force, a guardian of the flame,
With radiance and brilliance, you etch your cosmic name.
Your beauty lies within, a hidden treasure chest,
Unveiling mysteries, as secrets are confessed.

So let us celebrate, this element of might,
Zirconium, the silent hero, shining through the night.
A testament to strength, and nature's grand design,
In your atomic heart, the wonders of science align.

TWENTY-EIGHT

JEWEL OF SCIENCE

In the cosmic dance of the periodic table,
Amidst the stars and the celestial fable,
Lies an element, both strong and rare,
Zirconium, a gem beyond compare.

Born in the heart of a fiery star,
Forged in the depths, from afar,
Zirconium, oh noble and true,
With properties that astound and woo.

A lustrous metal, silver and bright,
Resistant to corrosion, a shining knight,
In alloys and ceramics, it finds its place,
Enhancing strength with elegance and grace.

From nuclear reactors to satellites in space,
Zirconium, a guardian, holds its embrace,

With a melting heart, it radiates heat,
Enduring the elements, a sturdy feat.
    A jewel of science, with secrets untold,
Zirconium, a marvel, it unfolds,
In its crystalline lattice, beauty lies,
A tapestry of atoms, a cosmic prize.
    So let us marvel at this element rare,
Zirconium, beyond compare,
A testament to nature's divine,
A symbol of strength, for all of time.

TWENTY-NINE

WONDROUS FABLE

In the realm of elements, Zirconium stands tall,
A shining star, majestic and strong in its call.
With atomic number forty, it holds its own place,
A metal so versatile, full of grace.

From the depths of the Earth, it is truly born,
A treasure unearthed, yet often overlooked and worn.
Its lustrous demeanor, a sight to behold,
A symbol of endurance, as legends have told.

Zirconium, oh Zirconium, a wonder of creation,
Your properties, a source of fascination.
With a melting point that defies the heat,
You withstand the flames, never to retreat.

In alloys and ceramics, you lend your might,
Enhancing their strength, shining so bright.

From spacecraft to nuclear reactors you reside,
A guardian of progress, always by our side.

But beyond your strength, Zirconium dear,
A secret lies within, crystal clear.
For in the hearts of lovers, you hold a key,
As a gemstone, adorning fingers, for all to see.

Zirconium, oh Zirconium, a jewel both fierce and tender,
Your presence, a reminder of love's splendor.
A symbol of devotion, forever entwined,
In circles unbroken, a love undefined.

So let us celebrate, this element rare,
With poems and facts, to show we care.
Zirconium, the unsung hero of the periodic table,
A testament to nature's wondrous fable.

THIRTY

PILLAR OF STRENGTH

In the realm of elements, a wonder lies,
Zirconium, a gem that captivates the skies.
A metal so noble, a symphony of grace,
Its secrets unravel with every embrace.

Born in the core of celestial fire,
Zirconium's tale, a cosmic desire.
Stardust and supernovas, its genesis true,
A bewitching fusion, ancient and new.

A phoenix reborn, in molten embrace,
Zirconium emerges with elegance and grace.
Its lustrous allure, a radiant glow,
A testament to beauty, a story to show.

In laboratories, its secrets unfold,
An alchemist's dream, a tale untold.

A catalyst of change, it weaves its spell,
Transforming the world, where wonders dwell.
    From aerospace giants to nuclear might,
Zirconium's essence, a guiding light.
Strengthening alloys, a pillar of strength,
Innovating horizons, pushing the length.
    A protector of vessels, a shield so bold,
Zirconium's armor, a tale to be told.
Resistant to corrosion, a guardian true,
Preserving the sanctity of what it comes through.
    Oh, Zirconium, a marvel divine,
Your presence, a treasure, forever to shine.
A symbol of resilience, a beacon of might,
In the tapestry of elements, your brilliance takes flight.

THIRTY-ONE

MYSTICAL PRIME

In the realm of fire and molten glow,
Where elements dance and secrets flow,
Lies a gem, a radiant sight,
Zirconium, shining with celestial light.

Born from the stars, in cosmic embrace,
Its essence pure, with elegance and grace,
A metal so rare, a treasure untold,
Zirconium, a story waiting to unfold.

In the depths of Earth, where magma churns,
Zirconium emerges, as the fire yearns,
A guardian of strength, a warrior bold,
Its presence shimmering, a tale yet untold.

With atomic prowess, it bonds with might,
A catalyst of change, igniting the night,

In laboratories, where wonders are found,
Zirconium whispers, with wisdom profound.

From ceramics to alloys, it weaves its spell,
Enhancing the world, where wonders dwell,
A protector of secrets, a keeper of time,
Zirconium shines, in its mystical prime.

Through centuries past, and futures unknown,
Zirconium's essence, forever has grown,
A gift from the cosmos, a jewel so rare,
Zirconium, a symbol of hope and flair.

So let us marvel, at this element divine,
Zirconium, a radiant star that will forever shine,
A testament to nature's grand design,
Zirconium, a legacy, forever entwined.

THIRTY-TWO

COSMIC TRAVELER

In the realm where fires bloom,
A metal gleams, Zirconium's tomb.
Born in blazing cosmic flame,
Its atomic dance, a radiant game.

A lustrous jewel, with strength untold,
Zirconium's beauty, a tale unfold.
A silvery shade, it adorns the sky,
Reflecting light, as clouds pass by.

On Earth it finds its humble place,
A stalwart guardian, full of grace.
In alloys, it lends its steadfast might,
Fusing worlds, in the dark of night.

A master of heat, it braves the fire,
Enduring tests, it does not tire.

From reactor's core to the surgeon's knife,
Zirconium's touch brings forth new life.

   Yet, beyond its earthly bounds it flies,
In distant stars, where our gaze lies.
A cosmic traveler, through space and time,
Zirconium, a celestial rhyme.

   Its bonds unbroken, in every form,
A testament to its enduring norm.
Zirconium, a symbol of resilience true,
A reminder of the strength in me and you.

   So let us cherish this metal rare,
A symbol of hope, beyond compare.
For in Zirconium's dance, we find,
A universe of wonders, intertwined.

THIRTY-THREE

A MASTERPIECE

In the realm of elements, Zirconium stands tall,
A silvery beauty, adorning the periodic hall.
With a shimmering gleam, it catches the eye,
A marvel of nature, reaching for the sky.

Born in the fiery depths of Earth's crust,
Zirconium emerges, a symbol of trust.
Strong and resilient, it withstands the test,
A guardian of structures, at its very best.

From nuclear reactors to spacecraft so grand,
Zirconium aids us, with a helping hand.
Its alloys forged, with brilliance and might,
Pushing boundaries, soaring to new heights.

But beyond its strength, lies a secret untold,
Zirconium whispers stories, precious and bold.
For within its core, a treasure does hide,
Radiant gems, where dreams coincide.

Zirconium, a custodian of dreams and desires,
In its crystal lattice, passion transpires.
A token of love, a symbol of devotion,
Enchanting hearts with a celestial motion.

So let us marvel at Zirconium's reign,
A luminary in science, it shall remain.
A testament to nature's boundless art,
Zirconium, a masterpiece, etched in our heart.

THIRTY-FOUR

SCIENTIFIC DELIGHT

In the depths of Earth's core,
Where molten fires roar,
Lies a gem, a precious stone,
Zirconium, its beauty unknown.

A metal strong, yet light in weight,
It weaves its magic, an alloy's fate.
Silvery-white, with lustrous sheen,
Zirconium, a wonder seen.

In the laboratory's gentle flame,
It dances, igniting a vibrant game.
From zircon sand, it's skillfully derived,
A creation that leaves us all surprised.

In the realm of science, it finds its place,
With properties that time cannot erase.

Resistant to corrosion, a warrior true,
Zirconium, a guardian through and through.

In nuclear reactors, it stands its ground,
Shielding us from danger, profound.
Its strength and endurance unsurpassed,
Zirconium, a metal built to last.

Yet beyond its strength, a hidden charm,
A secret that sets it apart from the swarm.
For within its crystal lattice, deep within,
Lies a tale of history, a story to begin.

Zirconium, a witness to ages gone by,
A relic of Earth's ancient sigh.
A testament to time's enduring grace,
Zirconium, a jewel in time's embrace.

So let us marvel at this element rare,
With its secrets and wonders beyond compare.
Zirconium, a gem of scientific delight,
Forever shining, in the realm of light.

THIRTY-FIVE

CATALYST OF CHANGE

In the depths of the earth, a hidden treasure lies,
A shimmering gem, where beauty truly lies.
Zirconium, the element of fire and grace,
With secrets untold, bound to amaze.

A metal so strong, yet delicate in form,
Zirconium, the catalyst of a celestial storm.
A lustrous creation, adorning the hands of time,
Its essence enduring, forever sublime.

In the heart of stars, its birth did occur,
Nurtured by cosmic forces, pure and pure.
Emerging from the cosmic forge's embrace,
Zirconium, a cosmic gift, in its rightful place.

Its atomic dance, a symphony unseen,
Unveiling the mysteries, like a dream.
The chemistry of life, intertwined and entwined,
Zirconium, the element, forever enshrined.

In laboratories, its wonders unfold,
Advancing our knowledge, stories untold.
A bridge between worlds, a catalyst of change,
Zirconium, the element, our constant exchange.

From jewelry to reactors, its purpose diverse,
Zirconium, the element, a gift to the universe.
With elegance and strength, it shines so bright,
A symbol of endurance, a guiding light.

Oh, Zirconium, element of dreams,
A tapestry woven in celestial seams.
In your presence, we find awe and delight,
A testament to the wonders of the night.

ABOUT THE AUTHOR

Walter the Educator is one of the pseudonyms for Walter Anderson. Formally educated in Chemistry, Business, and Education, he is an educator, an author, a diverse entrepreneur, and he is the son of a disabled war veteran. "Walter the Educator" shares his time between educating and creating. He holds interests and owns several creative projects that entertain, enlighten, enhance, and educate, hoping to inspire and motivate you.

> Follow, find new works, and stay up to date
> with Walter the Educator™
> at WaltertheEducator.com

www.ingramcontent.com/pod-product-compliance
Lightning Source LLC
LaVergne TN
LVHW052000060526
838201LV00059B/3747